TULSA CITY-COUNTY LIBRARY

SCJC

S0-ANE-156

ALIENS IN THE MOVIES

by Jenny Mason

STORY LIBRARY
MORE TO EXPLORE

www.12StoryLibrary.com

12-Story Library is an imprint of Bookstaves.

Copyright © 2022 by 12-Story Library, Mankato, MN 56002. All rights reserved. No part of this book may be reproduced or utilized in any form or by any means without written permission from the publisher.

Developed and produced for 12-Story Library by Focus Strategic Communications Inc.

Library of Congress Cataloging-in-Publication Data
Names: Mason, Jenny (Children's author), author.
Title: Aliens in the movies / by Jenny Mason.
Description: Mankato, Minnesota : 12-Story Library, [2022] | Series: Aliens! |
Includes bibliographical references and index. | Audience: Ages 10–13 | Audience: Grades 4–6
Identifiers: LCCN 2020014846 (print) | LCCN 2020014847 (ebook) | ISBN 9781632359339 (library binding) |
ISBN 9781632359681 (paperback) | ISBN 9781645820994 (pdf)
Subjects: LCSH: Extraterrestrial beings in motion pictures—Juvenile literature.
Classification: LCC QB54 .M374 2022 (print) | LCC QB54 (ebook) | DDC 791.43/615—dc23
LC record available at https://lccn.loc.gov/2020014846
LC ebook record available at https://lccn.loc.gov/2020014847

Photographs ©: GrandeDuc/Shutterstock.com, cover; Allstar Picture Library/Alamy, 1; PD, 2; Georges Méliès/PD, 5; Liu zishan/Shutterstock.com, 5; Allstar Picture Library/Alamy, 6; AF archive/Alamy, 7; NASA on The Commons, 7; Sportsphoto/Alamy, 8; AF Archive/Alamy, 9; Sportsphoto/Alamy, 9; Allstar Picture Library/Alamy, 10; Sitthipong Pengjan/Shutterstock.com, 11; XX, 11, XX, 11; Allstar Picture Library/Alamy, 12; dpa picture alliance/Alamy, 12; mmckinneyphotography/Shutterstock.com, 13; PD, 13; Castleski/Shutterstock.com, 14; Tinseltown/Shutterstock.com, 15; samzsolti/Shutterstock.com, 15; Nikolas Coukouma/CC2.5, 16; PD, 16; The Image Worx/Shutterstock.com, 17; XX, 17; PD, 17; Russell Hart/Alamy, 18; betto rodriques/Shutterstock.com, 19; Atlaspix/Alamy, 19; Atlaspix/Alamy, 20; AF Archive/Alamy, 20; Sportsphoto/Alamy, 21; Heritage Image Partnership Ltd./Alamy, 22; History and Art Collection/Alamy, 23; TCD/Prod.DB/Alamy, 23; Uncleleo/Shutterstock.com, 24; AF Archive/Alamy, 25; Yuri Turkov/Shutterstock.com, 25; RGR Collection/Alamy, 26; NASA, 27; PictureLux/The Hollywood Archive/Alamy, 28; Beth Madison/CC2.0, 29; Twocoms/Shutterstock.com, 29

About the Cover

Giant alien machines invade a city.

Access free, up-to-date content on this topic plus a full digital version of this book. Scan the QR code on page 31 or use your school's login at 12StoryLibrary.com.

Table of Contents

First of its Kind: *Trip to the Moon*

George Méliès, circa 1890.

In 1902, George Méliès did two things no moviemaker had done before. First, he made the first science fiction film. Second, he made the first alien movie.

The French director called the movie *Trip to the Moon*. Science fiction imagines how far technology can advance. It shows possible future worlds. *Trip to the Moon*

A scene from *Trip to the Moon*, 1902.

14

Length in minutes of *Trip to the Moon*

- The movie takes up 825 feet (251 m) of film tape. Movies today are digitally recorded.
- Méliès spent 10,000 francs on the movie. That was a big budget then.
- The amount is equal to about $140,000 today.

imagined humankind exploring outer space. Space travel stories were popular in the 1800s and 1900s.

In Méliès's movie, astronomers and scientists fly a rocket ship to the moon. In an underground mushroom cave, they battle aliens called Selenites. The movie was an international hit.

DEEP DIVE

Not all alien movies occur in outer space. Director James Cameron went under water for his 1989 film *The Abyss*. People discover an alien city at the bottom of the ocean. The peaceful aliens try to prevent nuclear wars.

Queen of the Aliens: Zoe Saldana

Zoe Saldana (right) in *Avatar*.

Zoe Saldana rules outer space. At least in Hollywood. In 2009, she starred in *Avatar* as an alien. Her character, Neytiri, is a tall, blue-skinned extraterrestrial with a long tail. In *Guardians of the Galaxy*, she played Gamora, a green alien warrior.

In the latest *Star Trek* movies, Saldana played Nyota Uhura. Uhura is a human. She is an expert in alien languages.

Saldana grew up in the Dominican Republic. Her family moved to the US when she was 9. She learned ballet.

Zoe Saldana in *Avengers: Infinity War*, 2018.

In high school, she switched to theater. She made the leap to alien movies easily. She was a big science fiction fan. Her mom was a Trekkie. Trekkies are *Star Trek* fans.

Saldana is well on her way to becoming the top film actress in the industry. Her many space and alien movies are major hits around the world.

10
Neytiri's height in feet (3 m)

- *Avatar* was the world's most successful movie for almost 10 years.
- Actress Nichelle Nichols played Uhura in the 1960s TV version of *Star Trek*.
- Because of Uhura, Civil Rights leader Martin Luther King, Jr. was a *Star Trek* fan.

Nichelle Nichols.

Hopes and Fears: Symbolic Aliens

Aliens in movies come in all varieties. In any form, aliens are symbols. They represent what humans fear or wish for.

Superhero aliens are a great example. *Superman* (1978) and recent Marvel movies star aliens with mighty powers. Thor. Thanos. Loki. They all have powers people want.

In the 1950s, alien invasions were common movie plots. *The War of the Worlds* (1953) and *Invasion of the Body Snatchers* (1956) are perfect examples. In those films, aliens symbolized fears about war.

Aliens have also been symbols of technology, bad or good.

Christopher Reeve played Superman in the 1978 movie.

A battle scene from *The War of the Worlds*.

Some living alien machines are villains. The Chitauri in the Marvel movies are one case. Meanwhile, Autobots in the Transformers movies are heroic aliens.

FILMING FEAR

In *The Day the Earth Stood Still* (1951), an alien visits Washington, DC. He delivers a message. War is not welcome in space. The movie played on common fears about scary UFO sightings in 1947.

202

Number of other-world creatures actor Bill Blair has played

- Blair set the Guinness World Record. He is called the "Alien Actor."
- Tom Woodruff Jr. is known as a creature performer.
- He famously played the xenomorph in the third *Alien* movie (1992).

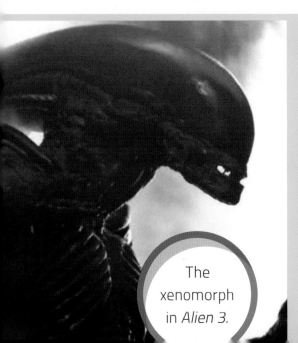

The xenomorph in *Alien 3*.

4

Out of This World: The Best Alien Movie Ever

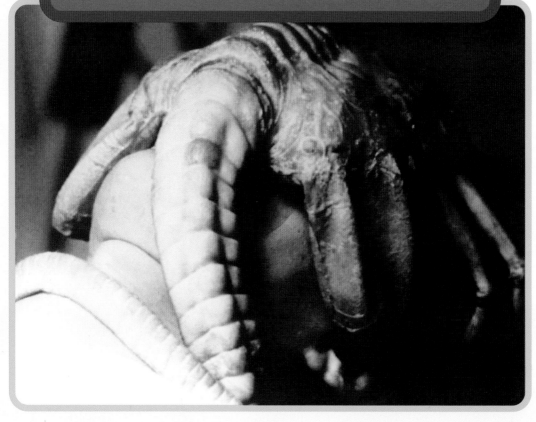

And the award for "Best Alien Movie Ever" goes to... Hang on. Choosing the "best" alien movie is a tricky challenge. In an online vote, movie fans chose the 1979 horror film *Alien*.

Should the "best" be decided by dollars? Marvel Universe movies have piled up $18 billion. In *Captain Marvel*, two alien races are at war. *Guardians of the Galaxy* features the alien Gamora.

She is the deadliest fighter in the galaxy.

But that is a whole series. The award is for a single movie, right? *Avatar* set a record, making $3 billion. Plus, *Avatar*'s main characters are aliens. Most Marvel characters are humans with super powers. Maybe the "best" alien movie is whichever one happens to be playing on your screen right now.

LARSON · JACKSON · MENDELSOHN · HOUNSOU · LYNCH · BENING · GREGG

19
Number of sensors in a motion capture suit, or *mocap*

- Actors in *Avatar* wore mocaps during filming.
- Computers turned their digital images into the tall, blue Na'vi aliens.
- The film mixes 40-percent live action with 60-percent computer-generated imagery (CGI).

THINK ABOUT IT

What is your pick for the best alien movie of all time? Why did you choose it?

5

In Other Words: The Rise of Klingon

Klingons are a tough alien race. They eat worms. They hate a soft bed. They want war, not peace. And humans have been trying to learn their language for decades.

Klingons come from the popular TV series *Star Trek*. In 1979, *Star Trek* became a movie.

Mark Okrand, 2015.

It featured a few invented Klingon words. As moviemakers made more films, fans wanted more Klingon. Marc Okrand was hired to create the alien language. He is a language expert.

Michael Dorn as a Klingon in the 1987 *Star Trek: The Next Generation* movie.

Invented languages are called *conlangs*. That is short for *constructed languages*. Many other alien films use conlangs. *Avatar* invented the Na'vi language. *Star Wars* created languages for Wookiees, Ewoks, and other beings. Many conlangs are based on real languages or sounds.

William Shakespeare's plays *Hamlet* and *Much Ado About Nothing* have been translated into Klingon. A Dutch company produced an entire opera in Klingon.

Wookiee language uses the sounds of animals such as bears, lions, and walruses.

THINK ABOUT IT

Besides words, what else makes a language? What are other ways people communicate?

STAR TREK
THE OFFICIAL GUIDE TO KLINGON WORDS AND PHRASES
By MARC OKRAND
THE **KLINGON DICTIONARY**
ENGLISH/KLINGON
KLINGON/ENGLISH

3,000
Total number of Klingon words

- English has 40,000 words. Klingons have no word for hello.
- The official Klingon dictionary has sold over 300,000 copies.
- Klingon is based on a lost Native American language.

6

Bursting at the Scenes: Xenomorph

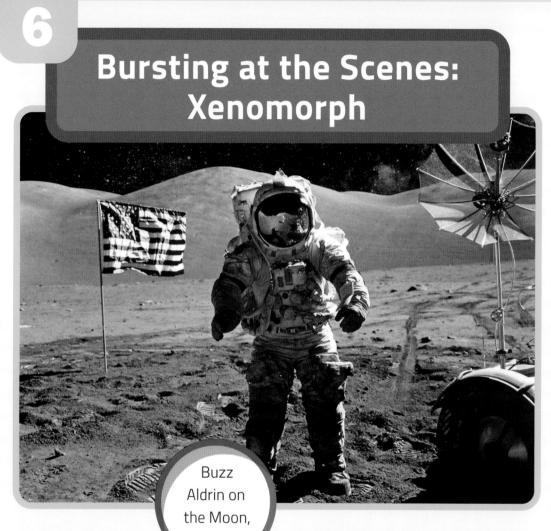

Buzz Aldrin on the Moon, 1969.

Alien and space travel movies surged after 1969. That year, American astronauts explored the Moon. In *Star Trek* and *Star Wars*, outer space was glamorous. Aliens were ordinary and mostly friendly. In 1979, one movie changed all that.

Alien was the first real horror film in space. Director Ridley Scott created an ugly outer space. In the story, a space mining crew was trapped in a spaceship with an alien monster. The alien was called a *xenomorph*. It was unlike anything audiences had ever seen.

14

The xenomorph body blended snake, sea creature, and black widow spider. It had two mouths. One bit. One stabbed victims. The xenomorph also laid eggs inside people. When they hatched, a new alien would burst from their chest. *Alien* forever changed aliens in Hollywood.

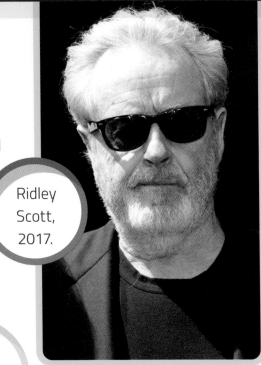

Ridley Scott, 2017.

Model of a xenomorph.

$1 billion

Profit from *Alien*, its movies, toys, and video games

- The idea for the xenomorph came from artwork by Swiss painter H. R. Giger.
- Giger grew up during World War II. His artwork spilled from his nightmares.
- The *Alien* movie received many awards in both the US and Britain.

7

In Worlds to Come: Diverse Aliens

Discovering new planets has created new sciences. New voices have joined the scientific conversation. The same is true in science fiction. Science fiction first imagined alien stories.

Historically, white men dominated this field in books and movies. More recently, female and non-white creators are blazing new paths. Octavia Butler is a key example. Butler's stories feature women of color.

Octavia Butler wrote the Xenogenesis series.

XENOGENESIS
by Octavia E. Butler

14

Percent of female main characters in science fiction movies by 2014

- Zoe Saldana was in three science fiction movies before 2014.
- Eight science fiction movies featured a non-white male main character. Will Smith starred in six.
- Writers N. K. Jemison, Ann Leckie, and Nnedi Okorafor are adding diversity to science fiction.

Will Smith, 2012.

She also created the Oankali aliens. They have three genders.

Butler's alien series will soon become a TV series. Perhaps Hollywood will adapt them into movies. Now is the time to create limitless futures.

Frankenstein
MARY SHELLEY

Frankenstein

ILLUSTRATED BY NINO CARBÉ

LADIES FIRST

Author Mary Shelley is credited for inventing today's science fiction. Her 1818 novel *Frankenstein; or, The Modern Prometheus*, shows the dark side of technology. Like Shelley, Ursula K. Le Guin is an influential female in science fiction. She imagined aliens without gender.

17

Looking Good:
Aliens in Makeup

Visual effects (VFX) teams turn fantasy into reality. Especially in alien movies. VFX artists use a variety of materials and techniques.

The *Star Wars* FX crew built small models of AT-AT walkers. Then they filmed these models in motion. Actors performed in front of a special screen displaying the AT-AT footage. The screen created an illusion. Tiny actors seemed to run from giant robot walkers.

To transform actors into bizarre aliens, VFX artists make foam rubber masks. The process begins with modeling clay. Artists shape the alien features. Horns. Flaps. Spikes.

AT-AT walkers from *The Empire Strikes Back*, 1980.

165

Temperature in degrees Fahrenheit (78° C) for baking foam rubber masks

- Masks need about six hours in the oven. Putting on a mask can take two to three hours.
- The mountains behind the AT-AT walkers are paintings.
- The snow they stomp through is baking soda.

Tentacles. They make a mold of the mask. Next, they pump foam rubber into the mold. The mold is then baked. Once cooled, the rubber mask is ready for the actor or actress.

VFX teams can also use computer-generated imagery, or CGI. This tool added space ships, mountains, forests, and oceans to the alien thriller *Arrival* (2016).

A scene from *Arrival.*

Spatial Relationships: Alien Buddies

Chewbacca (left) and Han Solo from the 1977 *Star Wars* movie.

Not all movie aliens are monsters or invaders. In fact, many of them play the most important role of all. A best friend. In any galaxy, no one comes between the *Star Wars* team Han Solo and Chewbacca. Chewie speaks in rumbling howls. Like any best friend, Han always understands.

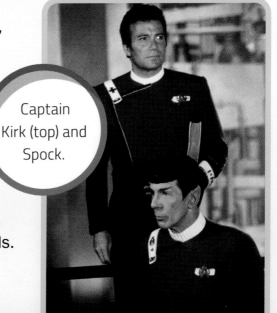

Captain Kirk (top) and Spock.

A scene from *Lilo and Stitch.*

Buddies are at the center of *Star Trek* movies and TV shows. Mr. Spock is Captain Kirk's best friend. Spock comes from the planet Vulcan. He is very logical. Kirk relies on Spock's wisdom.

In *Lilo and Stitch* (2002), Lilo teaches Stitch about friendship. Then in *Transformers* (2007), the alien robot Bumblebee befriends Sam Witwicky. Proof that some friends drive to the end of the universe and back.

1994

Year Steven Spielberg's *E.T. the Extra-Terrestrial* was preserved for the future

- The Library of Congress added this 1982 alien buddy story to its collection.
- From 1987 to 2005, Michael Westmore created every alien ever seen in a *Star Trek* movie or TV show.
- No one counted, but the total may be over a thousand creatures.

THINK ABOUT IT

What alien would you choose (or invent) to be your best friend? Why?

Over the Moon: How Aliens Travel from Books to Films

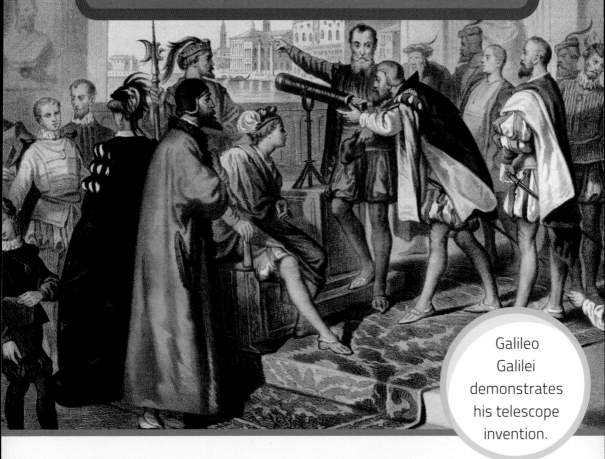

Galileo Galilei demonstrates his telescope invention.

On November 30, 1610, a full Moon changed the world. Galileo Galilei viewed the Moon through a telescope. He discovered worlds beyond Earth. The discovery sparked alien stories. Many are today's movies.

Cyrano de Bergerac wrote the first alien story in the 1600s. His funny adventure included Moon creatures. In the 1800s,

Cover of a 1927 magazine, featuring H.G. Wells's *War of the Worlds*.

MARS ATTACKS

H.G. Wells's novel *The War of the Worlds* was a hit in 1897. Forty years later, the story became a radio play. Just under a million listeners panicked. They thought the alien invasion was real.

Jules Verne and H.G. Wells wrote popular alien tales. Later, their books became movies.

12

Copies of the novel *Dune* in millions sold since it was first published in 1965

- Space travel, or astronautics, is the only science to begin as art.
- Space suits. Rockets. The launch countdown. Engineers took these ideas from books and art.
- The place where Jules Verne launched an imaginary rocket later became the Kennedy Space Center in Florida.

Today, many alien movies sprouted from books or comics. The list is very long. Frank Herbert's 1965 novel *Dune* is his most popular. It has several film versions. *Arrival* (2016) is also based on a book by Ted Chiang.

A scene from the 1984 movie *Dune*.

No Place Like Home: Filming on Alien Planets

Some movies create alien worlds using digital imagery. Others film "on location." That means they use real places. For example, Wadi Rum in the Jordan desert resembles Mars. It appears in *Red Planet* (2000), *The Last Days on Mars* (2013), and *The Rise of Skywalker* (2019).

Prometheus (2012) made Wadi Rum a different alien planet. The film crew also shot alien scenes beside a huge waterfall in Iceland.

Few films know the galaxy better than *Star Wars*. The desert planet Tatooine is actually Tunisia, Africa.

Wadi Rum.

Mark Hamill on an Irish island in *Star Wars: The Last Jedi*.

In *The Last Jedi* (2017), Luke Skywalker hides on an Irish island. *Star Wars* also used Seville, Spain, and a glacier in Norway.

500

Average number of people needed to make a movie

- Movies that film "on location" in foreign countries need bigger crews.
- Over 4,000 people helped with *Avengers: Infinity War* (2018). That movie features good and evil aliens.
- The visual effects (VFX) department needed over 2,100 people.

THE GREATEST TEACHER

The Jedi Master Yoda may be one of the most famous aliens. He appears in many *Star Wars* films. Speaks in a strange way, he does. He trains others to master the mysterious power called the Force.

25

Worthwhile Enterprise: More Diversity

women? Where were people of color from anywhere in the world? Were they stuck behind an alien cloaking shield?

Luckily, times have changed. The 2016 movie *Star Trek Beyond* has a very diverse cast. Female actors from the Dominican Republic and Algeria perform leading roles. One character is gay and is played by an Asian American. The original *Star Trek* TV series inspired this diversity in 1966.

Mr. Sulu was played by George Takei in the *Star Trek* TV series.

Gene Roddenberry created the show. He devoted it to diversity. As a result, the Starship Enterprise sailed space with a unique crew. It welcomed people of color. It welcomed people from different nations. It welcomed

For many decades, Hollywood movies had a problem. Important people of all kinds were missing. Where were powerful

The cast of the *Star Trek* TV series, 1976.

actors who were gay, lesbian, and gender-mixed. And yes, it welcomed aliens, too.

1992

Year Mae Jemison became the first African American woman astronaut in space

- The *Star Trek* character Nyota Uhura inspired Jemison.
- Hearing impaired. Blind. Differently abled. The show celebrates these differences in people and aliens.
- *Star Trek* kicked off 13 movies, cartoons, and other TV series.

Above and Beyond

We Are Family

Alien families are popular TV comedies. *3rd Rock from the Sun* followed an alien family from 1996 to 2001. They studied life on Earth.

The Neighbors placed a human family in an alien neighborhood. Alien families help audiences overcome stereotypical thinking.

Live Long and Prosper

Actor Leonard Nimoy played Mr. Spock on *Star Trek*. Spock is a world-famous alien. "Live long and prosper" was his motto. Nimoy taught people how to be better humans.

Just in Time

Dr. Who is a British television show. It stars an alien named "the Doctor." The show holds the Guinness Book of World Records for most popular science fiction TV show. It began in 1963 and is still running today.

Glossary

astronautics
The science and technology of human space travel.

astronomer
A scientist who studies planets, stars, and other bodies in space.

AT-AT walker
An all-terrain armored transport vehicle.

director
The person who supervises a movie or a play.

diverse
Something made up of people or things that are different from each other.

gender
a way of categorizing people, usually by male or female labels.

glamorous
Something that is exciting and attractive.

logical
Being very sensible, thinking clearly and carefully, using reason.

motto
A short expression or phrase that summarizes beliefs or goals.

mocap
The process of recording movements of objects or actors and using the information to create animated figures.

symbol
An action, object, or event that represents an idea.

Read More

Salisbury, Mark. *Moviemaking Magic of Star Wars: Creatures and Aliens*. New York, NY: Abrams Books, 2018.

Shetterly, Margot Lee. *Hidden Figures: Young Readers' Edition*. New York, NY: Harper Collins, 2016.

Suen, Anastasia. *Movie Props and Special Effects (Make It!)*. Vero Beach, FL: Rourke Educational Media, 2018.

Visit 12StoryLibrary.com

Scan the code or use your school's login at **12StoryLibrary.com** for recent updates about this topic and a full digital version of this book. Enjoy free access to:

- Digital ebook
- Breaking news updates
- Live content feeds
- Videos, interactive maps, and graphics
- Additional web resources

Note to educators: Visit 12StoryLibrary.com/register to sign up for free premium website access. Enjoy live content plus a full digital version of every 12-Story Library book you own for every student at your school.

Index

About the Author

Jenny Mason has always lived in UFO country. She grew up a few hours from Roswell, New Mexico. She currently lives in Colorado near the UFO Highway. She will go anywhere in search of amazing stories.

READ MORE FROM 12-STORY LIBRARY

Every 12-Story Library Book is available in many formats. For more information, visit **12StoryLibrary.com**